PLANT PARTS WE EAT

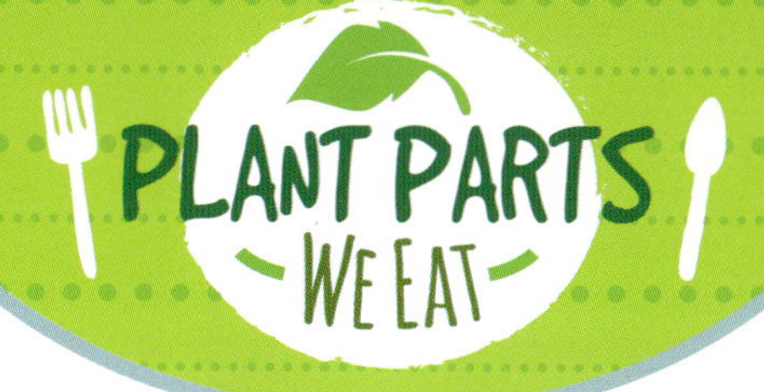

STEMS WE EAT

Katherine Rawson

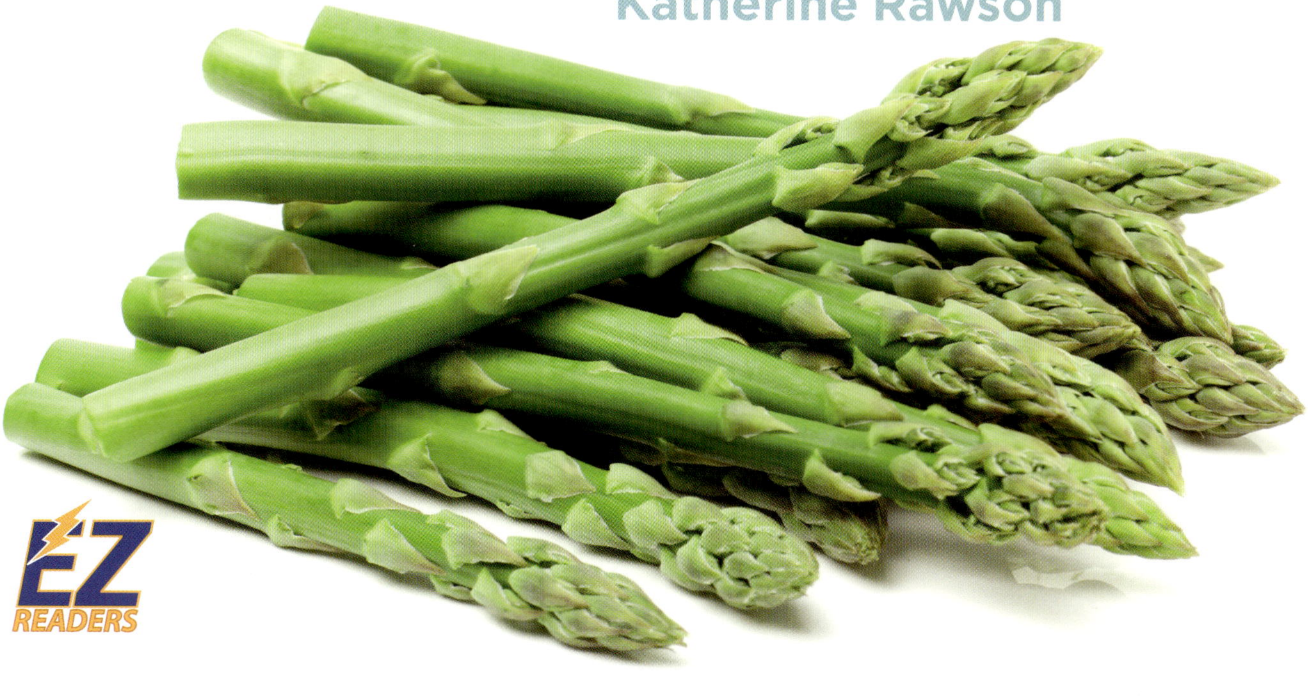

EZ READERS

Creating Young Nonfiction Readers

EZ Readers lets children delve into nonfiction at beginning reading levels. Young readers are introduced to new concepts, facts, ideas, and vocabulary.

Tips for Reading Nonfiction with Beginning Readers

Talk about Nonfiction
Begin by explaining that nonfiction books give us information that is true. The book will be organized around a specific topic or idea, and we may learn new facts through reading.

Look at the Parts
Most nonfiction books have helpful features. Our *EZ Readers* include a Contents page, an index, and color photographs. Share the purpose of these features with your reader.

Contents
Located at the front of a book, the Contents displays a list of the big ideas within the book and where to find them.

Index
An index is an alphabetical list of topics and the page numbers where they are found.

Photos/Charts
A lot of information can be found by "reading" the charts and photos found within nonfiction text. Help your reader learn more about the different ways information can be displayed.

With a little help and guidance about reading nonfiction, you can feel good about introducing a young reader to the world of *EZ Readers* nonfiction books.

Mitchell Lane
PUBLISHERS

2001 SW 31st Avenue
Hallandale, FL 33009
www.mitchelllane.com

First Edition, 2021.

Author: Katherine Rawson
Designer: Ed Morgan
Editor: Morgan Brody

Names/credits:
Title: Stems We Eat / by Katherine Rawson
Description: Hallandale, FL :
Mitchell Lane Publishers, [2021]

Series: Plant Parts We Eat
Library bound ISBN: 978-1-883845-05-6
eBook ISBN: 978-1-58415-103-6

EZ Readers is an imprint of Mitchell Lane Publishers.

Photo credits: Freepik.com, Shutterstock

Contents

Plants have stems.
The stems hold the plant up.

5

6

The stems carry water from the roots. They carry the water to the leaves. They carry **nutrients**, too.

We eat the stems of some plants. Celery **stalks** are stems.

We eat celery **raw**. It is crunchy. It tastes good with peanut butter. It tastes good in salads.

We eat celery cooked.
It tastes good in soups and sauces.
We can eat the leaves of celery, too.

Did You Know?

Celery leaves and seeds are used to flavor food.

Did You Know?

Purple asparagus turns green when cooked.

Asparagus is a stem. Asparagus can be green, purple, or white. It tastes good **steamed** or **roasted.**

Did You Know?

Asparagus needs sunlight to turn green. White asparagus is made by covering the asparagus in dirt as it grows.

Rhubarb is a stem.
Rhubarb tastes very sour.
Usually we cook it with sugar.
Then it tastes sweet.

Rhubarb tastes good in pies.
It tastes good in cakes.

Did You Know?

The stems of rhubarb are delicious, but the leaves are poisonous. Don't eat them!

Stems are good to eat!

Glossary

nutrients
Contents of food that we need for good health

raw
Not cooked

roasted
Cooked in the oven

stalks
Stems

steamed
Cooked in steam

Sources

https://www.britannica.com/plant/celery

https://www.britannica.com/science/stem-plant

https://www.cropsreview.com/plant-stem.html

http://www.wafarmtoschool.org/ToolKit/32/asparagus/Facts

https://www.gardeningknowhow.com/edible/vegetables/asparagus/varieties-of-asparagus.htm

https://www.csmonitor.com/The-Culture/Gardening/2009/0909/the-science-behind-purple-beans

https://extension.oregonstate.edu/food/safety-storage/are-rhubarb-leaves-toxic

https://homeguides.sfgate.com/parts-celery-plant-56996.html

https://allspiceonline.com/celery-leaf

Further Reading

Web Pages:
Read more about stems:
https://extension.illinois.edu/gpe/case1/c1facts2b.html

Learn how water and nutrients travel through a celery stem:
https://www.acs.org/content/dam/acsorg/education/resources/k-8/science-activities/motionenergy/graphing/celery-soaks-it-up-science-for-kids.pdf

Books:
How Plants Grow
by Dona Herweck Rice
(Teacher Created Materials, 2013)

The Amazing Life Cycle of Plants
by Kay Barnham
(B.E.S. Publishing, 2018)

Index

About the Author

Katherine Rawson loves growing, cooking, and eating vegetables. She also loves writing. So, she thought it would be a great idea to write books about plants we eat. Her favorite stem vegetable is rhubarb. She loves eating it for breakfast in the spring.